MATTER, LIFE

and

EVOLUTION

BY

JOHN G. ELLIOTT, D.A.

$4.50 Per Copy in Paperback
$6.95 Per Copy in Hard Cover
Add 60¢ for Postage and Handling

Available by Mail From

GIBSON-HILLER CO.
Publishers
P.O. Box 22
Dayton, Ohio 45406

Library of Congress Catalog Card Number 82-80274

ISBN 0-918892-04-X Paperback
ISBN 0-918892-03-1 Hard Cover

Printed by
McNaughton and Gunn, Inc.

Manufactured in the
United States of America

Table of Contents

Introduction

While science has been subdividing into a large number of specialties, it has also tended toward unification around a small number of themes. The physicists have made large strides toward the unified theory of all the forces of nature.

Philosophers throughout history have said that man is more than his material body. These past philosophers, however could not delve into the waves, fields of force, and particles of matter in order to know what was being sought.

The sum and substance of everything is consciousness. It is made up of attributes, of which man is but one. It would include a tree or any other inanimate objects along with such intangibles as honor, compassion, imagination, etc.

This work will also examine man's relations with himself and with others. It will show that life on earth is tied to the earth portion of the universe by bio-electromagnetic and bio-nuclear forces, and, that personal control of life will be increased through a better understanding of this part of the cosmic environment.

This material should stimulate the reader and produce ideas of value and meaning. It could also be helpful in thinking through many of life's problems.

John Gibson Elliott, D.A.

Chapter I

Basic Matter

There are many different particles that make up matter other than the electrons, photons, neutrons, etc. Many of them are even more elementary. These particles are governed by gravitation and electromagnetic fields, along with what is called the strong and weak fields found in the interior of the compounds that are called atoms. The quark is one of the elements of the atom compound. These quarks when close together have little attraction, however, when separated the least bit their magnetic attraction increases. These quarks have different electrical charges that determine how they will combine. These different charged quarks have each been assigned a color. This quark force theory is called quantum chromodynamics and is part of the Grand Unification theory that links elementary particles to the entire universe.

The chemical process used by thought to make itself aware is closer to being understood due to a new technique using an instrument called the laser. This laser fluorescence method is based on the knowledge that molecules absorb magnetic light waves of one frequency and then emit a different frequency. Thus, a substance as small as a single particle among several billion can be detected and identified.

Electromagnetism is an electric charge that gives rise to magnetism and is developed from the motion of electrons. In large bodies, such as the earth this force is called gravity. These two forces electromagnetism and gravity control much of the movements in and of the universe.

The electromagnetic force is made up of frequencies reading from single digit cycles to billions per second. The higher the frequency the shorter the wave pulse from each cycle. Only a small portion of this electromagnetic spectrum can be seen or felt by man's physical awareness system. Life, however is affected by much of this electromagnetic spectrum.

Part of this electromagnetic cosmic force comes to earth from pulsars and quasars. These dense neutron stars are millions of times the size of our sun. They are a long way off, being over a thousand million light years away. The entire universe is an electromagnetic system, of which the earth is a very small part.

The solar wind from our sun keeps much of this radiation from reaching the earth in lethal amounts. This solar wind originates in the sun's corona, and is made up of high energy particles that vary in amount, thus the amount of cosmic radiation received is not constant.

Space is not a vacuum, as it is filled with protons, electrons, and other particles, as well as all sorts of electromagnetic fields, waves of many lengths and strengths. Later in this work, space will be shown to be the essence of conductivity.

Solar flairs of our sun are gaseous in nature and are thousands of miles in diameter. The flares vary in both intensity and output of x-rays and light of both visible and invisible frequencies. These electromagnetic solar flares develop after sun spots and may last for several weeks.

Man is a bio-electromagnetic entity, much like the universe of which he is a part. Man radiates electromagnetic waves, some of which can be seen or measured, such as the aura and the brain wave. The effects of this electromagnetic radiation on life processes is now being used in many ways. The brain is constantly responding to waves and other forces from sources other than those produced on earth. The activity of the brain can be altered by this electromagnetic microclimate. One way it can do this is by altering its rhythm.

Earth life has been part of and subject to the cycles, rhythms, and waves from these external forces during its entire evolutionary development. That part of life called man, can and does vary the strength and effect of these forces. It is done through the use of biofeedback, transcendental meditation, and psychic self-regulation.

These, and other methods accomplish this by altering the biological phase-locked clocks that have been developed through long periods of cell evolution.

Self healing through the use of biofeedback is one of its many uses. There is also a close connection between biofeedback and such phenomena as clairvoyance, psychometry,telepathy, clairaudience and psychokiness. Through these methods the cells of body and mind can communicate with others. All living organisms possess an independent and organized voltage field that is part of the second and permanent body of that organism.

The sun through its cyclic electrical and particle emissions is a powerful factor in the development of events on this earth and other planets. Sun light and life are interrelated. Its rays serve many biological functions through its infrared and ultraviolet wave lengths. In photosynthesis sunlight is transformed into chemical energy, enabling growth of vegetation and in turn the vegetation changes carbon dioxide into oxygen.

Biological resonance and the interconnected electromagnetic system influence the make up of a large part of all life. Bioresonators, an element of the cytoplasm, play a roll in cell division and also as a link between cell growth and light. Thus by modulation of certain electromagnetic waves of light information is carried to the DNA and through its coil and membranes, that are connected in parallel in such a way that, a resonant circuit effect is produced. Thus cell activity is affected.

The study of the basis of matter and its quantized energy states is called quantum mechanics. Consciousness is the link between matter and the state of matter called life. The composition of consciousness is yet to be determined.

Conception begins with a single fertilized cell that contains all the DNA, half male, half female. By a process of dividing and redividing it stays in all the cells. The cells position themselves in a certain part of an organ or other structure, by the use of a polar coordinate system made up of receptor molecules. These molecules are located on or near the surface of each cell, and are lined with electrically active islands that recognize and join the receptors of similar cells.

From the small to the large, through Earth Science, more is being found and understood about the matter upon which all physical life operates. The entire surface of the earth is made up of thin rigid plates. These plates travel at a rate of a few inches a year. They are made of material from the interior of the earth. Most of this plate formation takes place along the earth circling ridges on the ocean floor. These moving plates cause volcanoes, mountains, and quakes. This movement over millions of years explains prehistoric differences and likenesses between forms etc. on the several continents.

Also, the earth has reversed its magnetic field several times. This is recorded in the plate material of the different ages.

Chapter II

Some Vital Qualities Explored

Incentives have a basic effect on the strength of desire. A good example of an incentive is advertising. Incentives deal with emotions and are given a vitality according to the degree of desire that is generated in the individual.

The degree of desire is also determined by the need of man's vital quality, and its demand for fulfillment. This degree of desire is a safe method of fulfillment were it not for certain incentives generated in the commercial world.

The use of initial effort starts an automatic action limited only by the nature of the thing that is acting. Every effort that man makes is aided by this automatic action. The amount of this automatic action depends upon the vitality of the initial lift and the amount of external opposition it meets.

A requisite for maintaining the strong vital quality of man is a proper elimination of things that no longer can be used to advantage. This elimination is mostly mental. Man has the ability to ignore the impressions continually pressing on his awareness. Thus things not given attention do not have a point of contact. Ideas and ideals are usually the property of the social order and are difficult to eliminate. In developing this ability small things should be worked on first.

The vital quality of imagination is a good tool to use in the process of elimination. It is a creative force that aids man in knowing what is needed and what can be disposed of. With this tool of imagination there is no need of straining toward self improvement through the use of will power.

Elimination, looked at in the larger view is best accomplished by having ready a new idea, plan, or action to replace the one to be eliminated. Man, by finding out what is wanted and concentrating on it, will by automatic action, remove unrelated thoughts, actions, and things from attention. What is not accepted or considered a necessary part of living for that degree of man, just does not exist.

Insulation, along with imagination and elimination, is also a good tool of quality. The use of this tool requires an effective and positive discriminating maintenance.

Every man, thought, force, and thing has power within the limits of its area or zone of action. These things cannot be avoided unless they are understood, so that their zones of action may be avoided, neutralized, or circumscribed. Thus the understanding, of the nature of things that are hindrances, makes possible a form of insulation that works. With this insulation, degrees of quality free themselves from belittling ideas and actions. This freeing is what makes room and time for the necessary and desirable activities of the degree's quality.

Vitality is measured by the impulse of desire. The impetus of desire depends on the strength of the impulse. The vitality of a race depends on the strength of its desire and aspiration as a group. The same principle applies to individuals. The impulse is increased by need or lack of some substance or quality, it is the driving force toward growth. Fate is the spending of impetus. Free will overcomes fate when fresh desires are impulsed sufficiently to override old desires.

It is by uniting intellect and intuition that man moves to larger or better goals. These two qualities when united give clarity of aim, and force of eagerness. Work on details that lead to improvement is done by the intellect, while the intuition is used to range ahead and chart the course toward the new goal. Along with these qualities of intellect and intuition goes the automatic action that results when the vital quality receives a constructive and purposeful impulse.

Prophecies are extensions of circumstance as known at the present time. If free will intervenes with new conditions, the prophecy is wrong. Man, individually, does not need to keep up with or master all the complexity of the modern world. Man needs only such details, as are usable in the furtherance of vitality and aspiration. These qualities act as a magnet in drawing the power of automatic action. This automatic action brings about outside conditions to match the reality that is seeking to be expressed.

It is only after this match of reality to aspiration, that the power of automatic action is achieved. Man then is able to profitably use the quality called intellect for the purpose of working over the details of the conditions and materials placed ready for use.

The use of many qualities as tools are needed in this process of materialization of idea, desire, or known lack. Science has found that there is a power in rhythm, and that strain or tension reduces the effectiveness. It is also known that the action or work must satisfy or be some type of enjoyment, in order to keep its power or vitality.

Stability is another essential quality of man. This tool is not far along in the development stage. It must be worked with in an easy fashion and in a planned way. The plan must be natural and not forced. Stability must be remembered to be a tool towards an end, not an end in its self.

The essential qualities make up the degree of awareness of the body, mind, and vital force.

These qualities man must work through in order to get the correct proportions for a plan of action. The plan's details now are ready to be worked on with joy and understanding, without dull depressing drudgery. Also much help will be received through the force of automatic action.

The essential qualities make up man's intangible portion of his organization. The degree of man depends on each individual's development of these essential qualities.

Chapter III

Toward Defining a Conviction About The Cosmos

Man explains the universe and his place in it. He makes rational, its laws and the direction of its changes. His answers are recorded in science, religion, and the arts. Further answers are added to the old field of knowledge. The addition of new knowledge calls for a shift in position.

The increase in new knowledge has been so great that man's shift in position has moved him almost totally out of his old position.

Through the use of the telescope, spectroscope, and the electron microscope things were found to be not as they were thought to be.

Philosophic systems attempt to answer questions. Each views the truth from a different angle, ignoring many other angles that have to be taken into consideration for a more correct answer. There is no ultimate knowledge at present to answer the questions about what is life or consciousness. However, relative knowledge has been increased from the simple to the complex.

This is seen clearly in evolution from the single cell to man.

The concept that all things have consciousness and are to some degree alive is wholly a point of view and definition. This angle of philosophic approach to life and consciousness has enlarged the threshold of known consciousness.

Consciousness is made up of all inanimate and animate structures. Both are composed of electrons, protons and photons etc., and respond to stimulus. The means to receive and respond to stimulus is the degree of consciousness present.

Man is becoming aware of higher forms of consciousness other than his own. Also, it is being determined that matter and life are elementally one and the same, however different in quality and degree.

Man's quality is capable of taking this abstract hypothesis and investigating how the accepted facts of life are explained by it. Matter and life is expressible, one in the terms of the other. Electronically matter is a balance of forces. Philosophically speaking matter is a state of consciousness. However, what is important is that matter, life, and consciousness may be manifested in each other as well as defined in terms of one another.

Evolution comes about in conformity to the idea which both precedes and governs development from within. The idea or intention is existent in all matter whatever its degree of quality. Idea, intention, and natural law all have intelligence. It is due to this essential characteristic of all matter, that man is able to utilize and rely upon it to get results.

All manifestation is preceded and governed by idea, intention, and force of life. Life is in many and varied forms, each made according to its own idea. Consciousness itself has many variations of quality. It is the appearance in substance of these differing qualities of consciousness that makes the diversity what it is.

Natural law works automatically when the proper conditions are present. This action is also the same for any quality of consciousness. Thus interplay and interdependence of life, force, and consciousness are always present and waiting for the proper conditions.

Individual manifestations of self-awareness are developed and expressed in new and more refined forms by the gathering of experience and memory. Memory and impression gathered from experience are stored in the quality of consciousness of the individual. It is by this action that the idea of consciousness is modified and improved. This is one of the ways that evolution works.

Instinct and racial wisdom are the result of the memory stored in consciousness, and are working toward the perfection of the idea, which that instinct or racial wisdom contains. The matter that embodies consciousness can be called the mechanism of awareness. It is also by memory that a faith is changed to knowledge.

Separate individual matter endures until it fulfills the idea of its consciousness, then it disassociates itself from the physical.

Matter, starting with the electron etc., gradually builds up into the one hundred plus elements, which in turn combine into substances. This evolution progresses in orderly fashion up to and including man. There are no short cuts known in this process from simplicity to complexity. Thus the duration of man and all other matter will last until the individual function is fulfilled. Thus is survival defined.

No individual matter, up to and including man can exist without being embodied in substance, and conditioned by space and time. All individual matter, visible and invisible is possessed of a body or substance thru which it can be aware.

Memory can be found in several forms. With the aid of the electron microscope, DNA and RNA were found to furnish one type. Another form of memory was discovered through the use of hypnotism. All memory is stored and will be found to be recoverable. No memory is ever lost.

Each degree of consciousness has its own fixed ratio of time to space, and its own general standard of measurement. Time and space are degrees of perception. To understand any other degree of consciousness, its time-space ratio must be manipulated until its concepts are understandable. One ratio cannot be applied directly to another. Also, the constant of other fields of consciousness as well as the ratio must be known before it is possible to enter into them.

Man's awareness mechanism is composed of physical sensation, instinctive reaction, intellectual processes of thought, intuition, and inspiration. They are all a part of consciousness. Just as white light, when it is broken down by a prism shows a series of colors, in an orderly fashion and continuous progression of wave length frequency, through out the entire spectrum. All the colors make up white light, just as all forms of awareness make up man's quality. It is just that present day man has not fully developed his entire mechanism of awareness. No awareness can stand alone. Awareness is the result, in direct lineage, of a long and unbroken series of contributing experiences and memories.

Free will is an exercise of choice. A creature merely reacts to stimulus. Man knows what he has chosen to do and that which he has chosen not to do. Man in constructing a course of conduct must keep this idea of free will constantly in mind. This choice should be made in a free moving and relaxed mode in harmony with the rhythm of nature.

Chapter IV
Awareness

Thought is a substance in the same sense as electricity is a substance. The mind tunes into this substance. Man can tune into different levels of thought forms according to his desires, needs, and degree of quality. Thoughts are tuned into or broadcast by attention and concentration. Desire and awareness of desire magnetically draws the corresponding thought forms to the receiving portion of the mind.

Conscious, voluntary contact with his field of thought forms increases the amount, strength, and clarity of the thought process.

Awareness of the vital force that is part of man, increases the ability to understand the reality of the universe. This vital force is a material thing and the use of this part of man is increased by development. Also this vital force is as real as any other physical substance. By understanding and developing the ability to use this vital force, man can move to a higher level of consciousness, comprehension, and awareness of what is important.

Creativity is one of the uses of man's vital force. The artist, the inventor, and the research people, etc. use it.

This force is sometimes called perception or inspiration and is separate from the lower level of man's intellect. This vital force can be used to find solutions to life's daily affairs.

The stimulus based on desire is self-originated and is a cause of mental and physical activity. The vitality of the desire or the clarity of perception determines the strength of the stimulus. Desire is the driving force of man and is separate from either the mind or the brain. It is part of the vital force of consciousness. New stimulus can override old, and in this way man can overcome destiny. Destiny is the spending of unarrested stimulus.

Awareness of control, and responsibility for actions or reactions activates the stimulus to revise or carry forward individual desires.

Each level of consciousness has prerogatives and capacity for growth to the next level. Each level must be completely understood before the next level is attained. These levels of awareness include all matter. This is another view of how evolution works.

Man, through awareness and study of natural laws, directs and uses them to further the assimilation of whatever portion or level of consciousness is at hand. The assimilation increases the ability and responsibility of each individual to move along the evolutionary path.

Awareness is gained through experimentation with

the natural laws and the vital force of the universe.

Man is clothed by, and works through an animal body with its limitations. The awareness of the process for growth increases as man understands his relation to the body.

Individual man raises the level of vital force by intentioning through desire and concentration. These forces are everywhere and are available by use of the proper magnetic attraction.

The absorption of these forces must be followed by their use. It is the natural law. The method of use depends on the level of the attraction. As a generator cannot store the substance called electricity, so the forces available to man, when induced, must be used as received or harm like that of lack of physical elimination, results.

Awareness of this automatic magnetic action activated by tuning into this all prevalent force of vitality gives abundant strength to whatever project is given attention. The project is the flow-through mechanism, as the motor or light is the flow-through mechanism for the generator.

Experimentation by man in the tuning into this force of vitality through concentration and inspiration has increased the capacity to receive, process, and expend it.

The substance called thought is part of this all

prevalent force. It is that portion of the force that is most used by man. The processing of this portion of the force, called thought, is a chemical action. The amount and type of chemical action depends upon the use that will be made of it. Awareness of the methods available for use of this field of force is dependent upon the level of consciousness.

Chapter V

Matter And Form

All matter in the universe has form. Form is made up by the consolidating action that takes place during the intermixture of time, space, and motion.

By changing the amount of time, space, or motion the form is altered. Also one form may exist inside another form as the seed in a fruit or the organs in an animal.

Invisible matter and force also have form. The invisible matter has more of motion than of time and space. Nothing can exist in the universe that does not have some of each of these three building blocks of the universe.

The entire universe is made up of time, space, and motion. The separation of visible and invisible matter is the frequency of the building block called motion. Also, the separation of different visible matter is governed by the mix of the building blocks. This also holds true for the invisible forms of matter, such as, light, thought, magnetic waves and the vital force of man, animals, and inert matter.

Each entity of matter, be it a stone, animal, magnetic wave or other force has form. The density of the form depends on its frequency. There are many views of the same entity. The black and white television picture is one view of matter with many of the frequencies screened out. Another view is obtained without screening out the color frequencies and is called color television. The strength of frequencies vary and this can be shown by an oscilloscope.

Consciousness is made up of a mixture of time, space, and frequency. The rock, bush, or man is made up of different mixtures of time, space, and frequency. Each mixture is set until changed by the slow process of evolution. This evolution is the change of the mix. Each degree of consciousness has a frequency band that contains sub-mixes. This accounts for the difference in flowers, animals, man, and rocks, etc.

There is only one consciousness in the universe. This consciousness has many forms made by the mix of time, space, and motion. This consciousness is governed by one law. This law may be called natural law, it is one and the same law for the universe.

This one law has many manifestations and extensions and is the basic law of the universe. This law governs all the mixes of time, space, and motion everywhere.

There are three dimensions in which all of the manifestations of consciousness operate. Consciousness operates only within the dimensions of time, space, and frequency. All matter in the universe, visible or invisible, is the result of the stress point developed by the mixture of time, space, and frequency. It is a unified field. This is how the law of creation works.

What consciousness consists of beyond its manifestations in the finite portion of the universe is beyond the finite mind of man to understand.

The high frequency, invisible form is always the container of the vital force. This form co-inhabits the low frequency, visible form. Upon separation, which causes the dissolution of the low frequency form, all experience and learning stays with the vital force in the high frequency invisible form.

This individual vital force in its high frequency form can again be attached to a low frequency form at a future date, for further experience and learning. This is another view of the working of evolution.

This individual high frequency form contains all of the senses used by the low frequency form. This high frequency form is also controlled by the vital force, which is also earth bound, due to its evolutionary status in the universe.

Communication between individual vital forces in the high frequency state is done in high frequency sound. The same goes for the other senses. Clothing is produced by the manipulation of frequency.

Both the visible and invisible forms have the solar

system as their natural habitat. The vital force of an individual present on earth must complete the evolutionary stages required for admission to the next step in the solar system.

The location or conditions of this next step are not known.

All laws operate for both forms of the vital force. The degree or status of the vital force is increased only by experience and learning. The higher the degree the greater the comprehension. The band of degree of consciousness for this solar system is the controlling element for the vital forces in it. Individual vital force due to the band of degree of consciousness, is limited to this solar system.

Each band of degree has a magnetic attraction that is the controlling factor. All that is connected with this solar system is in this degree band.

Through the evolutionary process, individual vital forces will graduate to the bottom of the next band of degree in another solar system.

Everything in the finite universe is in evolution. Everything includes this and the other galaxies. The law of evolution is all inclusive. A degree of consciousness in the infinite is beyond the ability of finite degree to comprehend.

Each band of degree has many sub-bands, such as mineral, vegetable, animal, and man. These sub bands are as separate as the main bands of the several solar systems. Also, each individual bit of consciousness varies in the sub-sub bands of degree frequency.

Gradual change in each band division is brought about by evolution. This evolution of vital force takes place in both the visible and invisible phases of the universe. There is just one set of laws for the entire universe. Man is learning more about laws that are known to exist. Man is also becoming dimly aware of many other existing laws.

Time, space, and frequency are elastic. Time is of short duration or long. Short, if the situation is pleasant. Long, if the situation is painful. Space is reduced if speed is increased. Motion or frequency may be increased or decreased. Since all matter has form and is contained by time, space, and frequency, it is also relative in its form. Man is able to produce change, to a certain degree, by the manipulation of time, space, and frequency.

Chapter VI

The High Frequency Phase

The individual entity enters the high frequency phase after so-called death. This phase is located in the same locality as the low frequency phase. Both phases are earth bound due to the magnetic field of this earth. There are exceptions, such as the visits to the moon, etc. Also entities evolve to other planets and solar systems in the finite portion of the universe.

In the high frequency phase man is active, orderly, and develops character. Each enitity has work that fits the capability. Each entity is met at the time of separation by an entity that speaks the language of the new arrival. Each newly arrived entity has an orientation period.

Each entity in the high frequency phase has an individual frequency, and that frequency is in the band that contains the other entities of the same general development.

Character development is the factor that places entities in the respective high frequency bands. Entities in a higher frequency band can enter a lower frequency band. However, lower band entities cannot enter higher frequency bands until further development permits it.

Desires, aspirations, and needs are available for fulfillment in each band. There is free and full communication between entities in each band.

Development of the entity goes on in both phases of life. Fulfillment of the requirements of a band in the high frequency phase moves the entity into the next band. Advancement, over many, many alternations between the low and high frequency phases of this earth, results in the transfer of the entity to another planet. This evolution goes on in all parts of the finite portion of the universe.

Grief, sorrow, melancholy is missing in the high frequency phase. They are experienced and must be outgrown in the low frequency phase.

Magnetic affinity in each band draws entities together for recognition and companionship. The entity is in contact with relatives, friends, and others in the frequency band.

Delving into this high frequency phase of life by low frequency entities is difficult and dangerous. A good chance of being obsessed is present. If the entity has a genuine gift or talent that provides a beat frequency that can be duplicated by an entity in the other phase, then communication can be established with safety. The entity with this ability is like a radio receiver. The entity must be able to tune out that material which is not wanted, and have firm control over that portion of the mental equipment that sets up the magnetic field in which the communication takes place.

Specific advice received by this type of communication should be received through entities that have a beat frequency high enough to screen out interference from entities of low intelligence and character. Communications received should not be acted upon without being checked out from every angle. Advice that tries to take the place of free will is to be rejected at once. Interference at all times should be guarded against by the use of good judgment and common sense. Conviction that it is possible to communicate with high frequency entities is dependent upon the development of the individual entity. The development of frequency is on an individual basis within the broad band of group development. Striving for growth and enlightenment is detrimental to development without the ingredients of experience, time, and assimilation etc.

Entities in the high frequency phase can aid entities in the low frequency phase only after the initial effort has been made. The initial effort sets up the magnetic attraction that can be complemented by the entities in the high frequency phase. The degree of help depends upon the degree of effort. Sustained effort brings sustained aid. Growth or advancement in the evolutionary process is due, in part, to this law of reciprocal aid. This law works in any field of endeavor.

Communications from the high frequency phase can come from entities in all stages of development, and mentality.

In passing into the high frequency phase, the entity brings all desires and accomplishments. Nothing is lost, with the exception of the low frequency form. Good or so-called evil entities are the same as they were in the low frequency phase. Caution and care must be practiced in choosing with whom one associates when communicating with entities that are in the higher frequency. If fine tuning of the beat frequency is maintained, entities of lower frequency have a more difficult time in entering the magnetic field of communication. Here again, common sense and good judgment enter the picture. Stability is lost if entities used forced growth methods in developing the beat frequency for intercommunication purposes. An entity can use any environment, good or bad, for the purpose of advancement. Entities should grow out of things. The forced denial of anything has no merit. The best effort put forth toward utilization of conditions and material at hand is the most productive way toward growth.

Chapter VII

Entity Development

Entities through functioning and self-development advance in the evolutionary process. The first and foremost duty of an entity is the building up of a solid nucleus of conviction that all entities are separate and immortal.

Whatever the entity develops above a basic need may be used to help other entities. Until basic needs are fulfilled any outgiving is both detrimental and unsafe. Full self-development, which is far above basic needs, takes innumerable repeated incarnations of the entity. The purpose of building the self is to gain ability to contribute help to other entities.

Methods to gain self-knowledge and control are through education, imagination, and acquired habits of right choice, etc. Each entity must find the best method towards this goal, for evolution hinges on the exercise of free will in the man degree of matter. Action or experience obtained by the entity is the way of evolution. The action or experience may be gained through the varied use of skills and understandings, that have been decided upon by the individual entity in the self-advancement process.

Desire used in conjunction with free will gives the entity more lift than will be achieved through the use of will power.

The substance that the entity is self-developing is enthusiasm, admiration, honor, humor, and many other aspects of the vital force. Help in the self-developing process is received from high frequency entities only after the effort toward development is started and sustained.

The aid received from the high frequency phase is many times in the form of inspiration. This inspiration is received after the effort has been put forth in the low frequency phase. The quality and amount of help received depends upon the strength of the magnetic attraction of the effort expended toward development. This is another way that evolution works.

It is necessary to have resistance to progress, that is why the effort is necessary. Without resistance, the tools called electronics and aircraft etc. would not work. The resistance may also be intellectual as well as mechanical. Creative imagination is the overcoming of resistance to change in method or idea. The frequency of an entity increases as progress is made in evolution. The difference between entities is frequency development. Different frequencies react to time and space, with results showing up as matter, animal, man, etc. Matter has a group entity. Animal and man have individual entities. This law of manifestation is the same as any other law. The law manifests automatically as a TV picture, sound, nuclear bomb, etc. when the conditions and amounts of time, space, and frequency are properly assembled and proportioned.

Man through law is self-developing. Also through the free will quality man is able to assemble and proportion the time, space, and frequency building blocks of the finite part of the universe. The assembly may be mechanical and the proportion of a chemical or intellectual nature.

Progress must start in the low frequency phase through the preparation of the proper magnetic attraction needed to draw aid from the high frequency phase of the finite universe. Awareness of this fact is a long stride forward in the self-development of the low frequency entity.

Vital current passes through an entity according to the development and strength of desire present. The specific form this current will manifest depends upon the entities purpose, intention, or idea.

The creative power of imagination uses the vital current to form what ever is within its field of magnetic attraction. Intelligence, through desire, can also use this current by magnetically tuning to it. The uses to which the intelligence can put this vital current also depends on the development of the entity.

Sickness is frequency out of tune. Each organ of an entity has its own frequency. Drugs are used to restore frequency. They release required frequencies, by chemical action, that restores the organ to its health frequency. Mental healing may reduce the use of drugs, if the proper frequency can be induced into the organ or other matter that is out of tune.

Service to other entities should only be given after the self has been completely filled by the vital current in the area that the service is to take place. Service to others should also be limited to the area in which accomplishment by the serving entity has been developed.

The intellect is a valuable tool for the entity and is the principle one now in use by man. The intellect works only when power is applied. The power is the vital current. This power is supplied to the intellect by desire, inspiration, and imagination, etc.

Law works through the conditions set up by the intelligence. The work is constructive or destructive, depending upon the intention of the entity. Law is the intention and will automatically bring about the event when conditions are correct. However law cannot remove the free will choice of the entity. The entity makes the choice, sets up the conditions that cause law to automatically act.

The low frequency body of the entity is the vehicle used by the mind to move the entity to places in which functioning is desired. The low frequency body ages. The high frequency body does not age and is the container of entity in both phases of life. The physical, low frequency phase, is renewed each time the entity returns from the high frequency phase. This alternation between phases continues for the entity until the intent of evolution is completed.

Chapter VIII
Functional Living

Man increases in knowledge and ability by doing. Each culture, by its doings, makes a contribution to following cultures in knowledge and ability. The main reason for life is evolution.

Doing is the process of development and growth. In the evolutionary sense, the doing is accompanied with eagerness based on a sincere sense of its importance toward producing something.

Dynamic functioning takes an idea and through the use of matter, puts the idea into form. In the process of giving form to an idea, man builds by the act of doing.

Vital force can only act during the functioning required to give form to an idea. The ability to function is determined by the stage of evolutionary development obtained by the entity. This stage of development is called character. Use is the end and aim for the development of anything.

Man is engaged in evolving. The process is cyclic in nature. The frequency of each period may be long or short.

The vital force is everywhere in the universe and is in constant use. It is used even to bend a finger or in thinking. The more use that is made of this vital force, the easier it is to use it. A large part of living is learning more and better ways to put this vital force to work. Each entity must find out the ways that are best suited to the entities present development.

Entities should live the low frequency phase of life to the fullest and with zest. This method of living will bring to the entity all the vital force needed, and without the necessity of using denial or withdrawal or other forms of asceticism.

Indulgence in negation and denial is as deterrent and destructive as any other form of devitalizing activity. Also, the entity will have to search for meanings and proportions for living that will produce the most utilization of this vital force.

The entity will change desires and actions as the character develops. Regret at the thought of giving up desires and actions just indicates that the entities development needs more time for preparation before change is undertaken.

Help to others should be given only from strength. Only, when the entity has full strength in the zone of action is the obligation to help present.

All entities in the low frequency phase can only express moral and ethical values through matter, both invisible and visible. Growth in any form, physical, mental, moral, and ethical can only be obtained through functioning.

An entity gains knowledge of anything only through working with it. If the entity has knowledge of something but has not worked with it in this session of the low frequency phase, then it was worked with in some previous session of the low frequency phase.

All skills, talents, and other aptitudes gained through work, will in time be used. The entity has an obligation to function with what has been developed. At the same time the entity may also be working at learning something new. The old and the new must be proportioned for the most gain.

Entities work best together when they are about equal in development. For whatever reason entities are gathered together, and the number of sessions spent in the low frequency phase, working toward understanding the problem of the group, is not about equal, there will be a lack of cooperation, affinity, or understanding.

Teachers and others in the school system see this lack in every forced grouping of students for the purpose of receiving instruction. A grouping on the basis of development and affinity would be much more productive.

An entity belongs to a family, a nation, a race, or any other group, with its advantages, limitations, and problems to be worked out because the group conditions are favorable for the entities functioning. An entity progresses in the evolutionary plan through functioning.

The struggle for existence and conflicts of interest are forms of resistance that are necessary during the process of development. In relation to the harmony of all, conflict between entities of a group will be resolved in favor of those whose actions favor the harmony of the group.

Entities when reaching out to function in new situations that are chosen through free will and desire may encounter defeat. The entity, however gains experience and wisdom through this new exploratory action. Also the experience gained may be used in later efforts to enter new fields.

The intellect functions through the free will mechanism in selecting the activity or direction that the entity will pursue, rather than selecting others that are offered. The adaptation of the entity to the new activity or direction is part of the process of evolution. Also, when a new direction or activity has been fully experienced through many sessions by entities in the low frequency phase, the activity or knowledge becomes part of the wisdom of the group.

Man should function without strain. If strain is encountered in the process of accomplishing a goal, the entity should through the free will mechanism review his desire in relation to the present power available for use. Power is a rhythmic substance that can be used to best advantage on the rising phase rather than on the ebb.

Progressions or evolution is rhythmic and life advances in waves. The entity will make the most progress with the least strain if this law is understood and followed.

The mechanism of free will must be used by the entity in the development of character. Character is the sum total of the entities functioning in the many sessions in the low frequency phase. Free choice is increased as character develops through the use of free will.

An entity is retarded, in the evolutionary sense, every time a decision is made affecting the entity either by another entity or even a machine, such as certain types of computers. It is mans privilege to make mistakes. Other forms of life are "locked in" by instinct, which is "Wisdom" perfected over millions of years of trial and error. The individual fly, bird, fish, etc. does not deviate from its group "wisdom."

Man through his ability to build and use complex tools, such as the electron microscope, is learning how the low frequency body is controlled. DNA and RNA are examples.

Performing suitable work against resistance develops the power for efficient evolutionary progress. Creative work is the process used for making out of matter, (thought is a form of matter), something that did not exist before. Man is continually working on new combinations of the vital frequency that makes up all matter and life.

No law pertaining to the universal vital frequency is made to work. The laws work automatically when the conditions or combinations are correct. Man discovered that this was so, through work done with chemicals and electronics.

The amount of this universal vital frequency that man can use depends upon the stage of development. The stage of development also indicates present character.

Beauty is a combination of the universal vital frequency that radiates and mixes in an harmonious way with the beholders reception mechanism. Beauty does not exist except in relation to the beholder. The sunrise or sunset was not beautiful to the ants or birds etc. before man came on the scene. Disharmony is incompletion in the thing beheld, or lack of development in the beholder, or both. The ugliness of disharmony and evil awaits the creative efforts of man to rework or complete the frequency combination, and so remove the disharmony.

Any learning by rote performance or forced growth has no permanent result. All permanent intellectual and moral powers are the result of growth. These powers earned by growth stay with the enitity when moving between phases.

If these ideas seem simple and true, to you, the reader, then you were ready for them.

Summary

The emphasis is on the oneness of consciousness. Each entity makes up a part of the whole, just as each link is part of a chain.

Entities that receive communications via the magnetic field, that is between the two phases of life, are called receiving stations, as in a closed circuit TV system. The entities on the high frequency side of life have the same senses and degree of development. The psychic gift is normal under natural law and may be developed just as any other talent. The reality of communication is in the subconscious part of the mind and on the right side of the brain. Successful communication depends partly on the degree of accuracy with which the subconscious mind registers the thought conveyed, and partly on the degree with which the thought passes out of the subconsciousness into the consciousness of the entity.

Coloring results when the conscious portion of the receiving station's mind overrules the subconscious with its own ideas. It is very difficult for a high frequency entity to pass a message through a receiving station without first suppressing many of the stations' subconscious memories. This suppression is necessary to insure the accuracy of the thought symbols being impressed on the subconscious of the receiving entity.

In the history of man there are two new ideas: One is evolution. This idea interprets the activities of life, such as art, industry, physics, chemistry, morality, etc. as ongoing expressions of consciousness. The second is the idea that there is only one set of laws for the universe.

The quality portion of consciousness, which is the character of the entity, is only developed in the high frequency phase. In the low frequency phase, the entity uses what additions to character were developed while in the high frequency phase during the latest trip taken there. Quality and quantity of consciousness is always in the forward direction for development. It is a one way street. Desire power is the instrument that motivates the entity to increase the quantity of character quality.

Force, matter, chemical reaction, and life itself are manifestations of degrees of consciousness. Creation is a development of higher and higher degrees of consciousness. From low degree, such as force and matter to high moral life, it is one and the same consciousness. Just degrees different due to many eons of qualitative and quantitative development, step by step from blind force up to reason, free will, and morality.

The freedom of perception, of thought, and of movement is acquired in relation to the degree of the entity. Every cycle of birth and rebirth wins the entity more understanding and therefore more freedom.

The high and the low frequency bodies that the entity posesses have of themselves distinct attributes of consciousness. Each organ and cell of the low frequency body (made up of its protons, photons, electrons, quarks, etc.) has a consciousness of its own.

Man on earth is made up of three parts. One part is the cellular, physical body. The second is the bio-electromagnetic plasma sometimes called the spirit body. The third part is the entity itself, often referred to as the I AM.

The entity always inhabits the bio-electromagnetic plasmic body, and the radiation from it is known as the aura.

Epilogue

The development and arrangement of life by influences, happenings and accomplishments for growth to an ultimate fullness is the purpose of living. There is an invisible inter-relation of all created things. This inter-relation will be better understood through an expansion of the consciousness. This absorption of comprehension is transmuted from the reality of the source into the symbols of the brain. The inner expansive faculty (electromagnetic) is the source that controls the physical apparatus called the brain. This inner faculty is the element used for communication via electromagnetic waves with other similar elements that are of the same frequency band. Telepathy is one example.

The more this inner faculty is developed the greater will be the creative ability. This faculty is an indestructible and also a universal force, always available for use when called upon. The more it is used, the more it can be used. This is best accomplished by expansion of desire rather than by brain concentration or the power of the will. This creative force is best used on mundane matters that have immediate usefulness. The brain will be further developed by its use.

While there is no substance in pure intellect, for it

is like a shadow on the wall, it is, however, when used by the brain, the creative gateway to actuality.

Service, duty, and will-power all create the dreariness of contraction and hinder the spontaneity of the body's functions. Desire power on the other hand gives a lift to happiness and harmony. It is like the hard work done in play. Very little is created under the lash.

Escape from the bodily restrictions of life can be experienced. The belief in the power to do it will expose (interface) the higher consciousness to the material brain. This belief gives a feeling and partial understanding of the next higher stratum of energy.

This change of mental environment takes place right here. Leaving the body and going away is only a matter of speaking. There is neither space nor distance to this higher stratum of consciousness, just greater revealing vision.

Space is a substance and depends on how it is looked upon and thought about. This area called space is dependent upon the degree of perception. The distance of the area is measured in the time it takes to cover it. A 186,030 mile walk takes a long time. By aircraft a shorter time. By electromagnetic wave it takes one second. Time and space are both relative.

A different center of the same consciousness is entered when the other and permanent body goes along with the entity to the conditions it was in prior to physical birth. This new center of consciousness has a resonance and an abounding energy along with an arresting beauty. Particulars about this higher level of consciousness are untranslatable to word symbols on the physical level of consciousness.

The knowledge of the true nature of spiritual practices has been distorted, dehumanized, scoffed at, misunderstood, and undesirably presented. Actually the inner life is to be desired and worked for, due to its beauty, warmth, and aliveness.

The exuberance of sinfulness can be lost if the unrelated spiritual example of trying to be good or trying to do good is entered into as a duty. The spiritual action must be enjoyable, natural, and warm hearted without the heavy hand of duty. The causal force of desire is part of the substance of reality. It can be gathered and tuned into for increased well-being.

Confidence is in itself a force. A definite belief in this force, gives the entity unfailing help in doing the things that are worth doing. This force is a part of the forward movement called evolution. Creative force is one of the vital parts of life. It is like an engine of power, dynamically creative, and capable of moulding the material world. Desire power used along with the essences of all the other qualities of life has compounded effects.

Consciousness, the vital principle, is the basis of all the manifestations and energies that make up the universe. It is the underlying evolutionary power, force, and life that keeps everything in development and functioning. The highest expression in this cellular portion of the universe is the human degree of consciousness.

The outflow of this possession of vital power is best motivated by desire. It makes little difference what it is used for, as it is the functioning that counts. As it is with electricity, it must flow to be useful.

Inspiration, also, like electricity must be brought into a condition that is ready for it. This part of the vital force is also undifferentiated and will work only when differentiated. The proper mind set is needed to develop the strength and ability to function. This mind set may come at odd and unexpected times.

Sincerity is needed for the accomplishment of creative work. Regulation and replenishment from within will assure that this form of the vital force will flow through with sufficient power.

When any force is generated, it must meet with an opposition. Opposition is itself a substance, and has a nature of its own. This opposition must be taken into consideration in the accomplishment of any creative work. Attention to the opposition should be paid to it, only for intellectual appraisement of its strength and to control its effect. This is accomplished by proceeding with the least friction and the most skill while working in the nature of the opposition. It is as the airplane rises with the least possible friction, on the resistance of the air.

Man's consciousness rises when dealing with provoking obstructions by thinking of them as a fun game, where you look beyond the opposition to success. It is working relaxed, confident, and with enthusiasm.

Glossary

Attribute

The qualities that make up the character of an entity, such as honor, mercy, honesty, etc..

Biofeedback

Knowledge of bodily processes needed by an entity in order to self manipulate, rather than be externally imposed upon by the uncontrolled natural flow through, of the forces of the universe.

Biological

Of or pertaining to life or living organisms.

Body

All life has two bodies. The cellular and the bio-electromagnetic plasmic. The respository of all experience and learning is the invisible, non-disintegratable body that is co-existent with the entity.

Cell

A microscopic and structured compound of mortal life.

Centimeter

0.01 meter or 0.393 inch.

Central Nervous System

That portion of the vertebrate nervous system, comprising the brain and spinal cord.

Circuit

The complete path traversed by an electric current.

Clairaudience

The act or power of hearing something not present but regarded as having objective reality.

Clairvoyance

The power of discerning objects not present to the senses, that have objective reality.

Consciousness

Self-awareness. The all-inclusive reality. Manifested in degrees, and is itself in evolution.

Corona

The outer-most solar atmosphere. It surrounds the super hot gaseous layer called the chromoshpere. This corona is at least two million kilometers from the sun itself. It becomes visible during a total eclipse.

Cosmic

The extraterrestrial regions of the universe.

Current

The net transfer of electric charge, or any other conserved, indestructible quantity across a surface per unit of time.

Cybernetics

The science of control and communication in all of their manifestations within and between machines, animals and man.

Cycle

Any series of changes or operations performed by or on a system that brings it back to its original state. Its general features include frequency, period, phase amplitude, and base level. It is found in biological and physical science activities and effects all matter and life.

Cytoplasm

The protoplasm of the cell external to the nucleus.

DNA

Deoxyribonucleic acid, a linear polymer made up of sugar 2-Deoxyribose, phosphate, and a purine base. It has a right handed helix structure which is encoded with genetic information.

Ecology

The study of the interrelationships which exist between organisms and their environment.

Electroencephalograph

An instrument used to record electric discharges of the cerebral cortex as detected by electrodes attached to the surface of the scalp.

Electromagnetic

The relationship of electric and magnetic fields in the propagation of waves and wave-particles called wavicles.

Electron

A stable elementary particle which is the negatively charged constituent of the atom.

Emission

Any radiation of energy by means of electromagnetic waves, as from the Sun or a radio transmitter.

Entrainment

The mixing of material into a preexisting organized material so that the material becomes part of the preexisting organized material.

Evolution

The processes of biological and organic change in organisms and other matter, by which descendants come to differ from their ancestors. In addition the earth-life phase is only half of evolution, and is called quantitative. The qualitative half of evolution is for character building. It is in the high frequency phase.

Field

The electric or magnetic force that exists between bodies not in contact.

Finite

Having definable limits or being susceptible of measurement in theory or in fact.

Frequency

The number of cycles, oscillations, or vibrations of a motion, wave or event. This repetition may or may not be regular and rhythmic.

Gluons

One of eight massless particles with spin quantum number and negative parity that mediate strong inter-actions between quarks.

Infinite

Inconceivable and immeasurable.

Kilometer

A unit of length equal to 1000 meters or to 1094 yards.

Laser

A narrow beam of coherent, powerful, and monochromatic electromagnetic radiation emitted by an electron device that excites the atoms of an optical resonator to an increased energy level and an in-phase radiation.

Light

Electromagnetic radiation of any wavelength. Its velocity is used in the theory of relativity as the connecting link between mass and energy.

Light Year

An astronomical measure of the distance a light ray covers in one calender year; about six to the twelfth power miles (6^{12}).

Macroscopic

Objects large enough to be studied or observed by the unaided eye.

Micro

A prefix indicating smallness and extreme sensitivity, as in microphone and microwave; approximately 10 to the negative six power (10^{-6}).

Microtubule

One of the hollow tubelike filaments found in cell components, such as cilia and mitotic spindle, and composed of repeating subunits of the protein tubulin.

Modulation

The process of varying the amplitude, frequency, or phase of a wave, or the velocity of an electron beam.

Molecule

The smallest unit of matter which can exist by itself and retain all of its chemical properties. It is a group of atoms held together by chemical forces.

Neutron

An elementary particle, without an electrical charge and with the same mass as a proton. Both are described as nucleons.

Nucleic Acid

A chainlike molecule of phosphoric acid, sugar, with a base of purine and pyrimidine. Examples are ribonucleic acid and deoxyribonulecic acid, as found in DNA and RNA.

Particle

A part of matter that is not a compound, such as an

electron or quark.

Phase-locked
: A system in which the frequencies of separate systems follow exactly the same phase.

Phenomena
: Any fact or event of unique significance susceptible of scientific description.

Photon
: A massless particle, the quantum of the electromagnetic field, carrying energy, momentum, and angular momentum.

Photosynthesis
: Synthesis of chemical compounds using light to manufacture carbohydrates from carbon dioxide and water, with a simultaneous liberation of oxygen, by chlorophyll-containing plants.

Physiological biophysics
: The use of physical mechanisms to explain the behavior and functioning and response of living organisms or their parts to physical forces.

Psychic
: Having the faculty of being sensitive to phenomena lying beyond the five senses.

Psychokinesis
: The ability to exert a mental influence on physical matter. Such as moving objects at a distance and bending metal etc.

Psychometry

Divination of facts about an object or its owner through contact with, or proximity to the object.

Proton

An elementary particle that is the positively charged constituent of ordinary matter and, along with the neutron is a building stone of all atomic nuclei.

Pulsar

A star source which emits radio frequency electromagnetic radiation in the form of pulses at regular intervals.

Quantum

One of the subranges of possible values of wave fields and the interaction of them with matter.

Quantum Chromodynamics (QCD)

The theory which predicts the behavior of quarks and their forces. Color instead of names is assigned to these new properties found in subatomic matter—hence chromodynamics.

Quantum Mechanics

Observes quantities of intensities and frequencies of spectral lines by dealing with their interactions.

Quark

A basic particle having a charge, with a magnitude of one third to two thirds that of an electron charge. They are found, for example in nucleons. They have a weak force close together, stronger as they get further apart. They combine according to their electrical charge, and are differentiated by color instead of names.

Quasar

An astronomical object, as far away as ten thousand light years. They are a quasi-stellar radiation source of great power.

Radiation

The process of emitting energy in the form of waves or particles through space or other medium.

Resonant circuit

An electric circuit adjusted to permit a maximum current flow when a voltage of a given cycles per second (CPS) is applied to it. Sometimes called a tank circuit.

Resonator

A device that indicates resonance at a particular frequency, such as a cavity resonator meter.

Rhythm

Movement marked by regular recurrence of elements, phenomena, etc,; hence periodicity.

RNA—Ribonucleic acid

A long chain of single stranded nucleic acid consisting of repeating units containing four heterocyclic organic bases: adenine, cytosine, quanine, and uracil. It transmits the coded information contained in the chromosomes of the nucleus of a cell to the protein-making ribosomes of the cytoplasm.

Solar flare
: An outburst of energy emission from an area in the region near a sunspot.

Solar wind
: The supersonic flow of gas, composed of ionized hydrogen and helium. It continuously flows from the sun out through the solar system. These high-energy particles are ejected from the Sun's outer layer or corona at speeds of about a million miles per hour, and at a temperature of a million degrees centigrade.

Space
: A three-dimensional entity that extends with unknown bounds. It is one of the three building blocks of the universe.

Spectrum
: An array of the components of an emission or wave separated and arranged in the order of a characteristic, such as wavelength, mass, or synergies, etc..

Spiritual
: Daily life carried on by a self, with higher associations and aspirations. Asceticism or other negations are something else.

Sun Spot

A temporary dark area in the photosphere of the sun, caused by a lowered surface temperature. It has eleven year cycle.

Tank circuit

A circuit which exhibits reasonance at one or more frequencies and which is capable of storing electric energy over a band of frequencies that are continuously distributed about the resonant frequency circuit, such as a coil and capacitor in parallel.

Time

Time has three aspects. Clock time which depends on the movement of the planet. Psychological time is the inner sense of time and is electromagnetic in nature and is controlled by forces entering the earth air space. The third time is that ever present time without start or end and is called infinite time.

Universe

Includes everything that can be known by the senses and their mechanical and chemical extensions along with all that is beyond them.

Telepathy

The direct communication between two minds, using electromagnetic waves.

Thought

This substance is much like electricity. It has its frequency band and form. It is also an attribute of consciousness and is a creative agent.

Wave

A disturbance which propagates from one point in a medium to other points without giving the medium as a whole any permanent displacement.

X-Ray

A penetrating electromagnetic radiation generated by high velocity electrons suddenly stopped by collision with a solid body.

Recommended Reading

Barnothy, M.F., Biological Effects of Magnetic Fields, Plenum Press, N.Y., 1969.

Becker, R.O., Electromagnetic Forces and Life Processes, Technology Review, December, 1972.

DeChardin, Pierre Teihard, The Phenomenon of Man, Harper and Brothers, N.Y., 1959.

Dewey, E.R., Cycles. The Mysterious Forces that Trigger Events, Manor Books, N.Y., 1973.

Dott, Robert H., Evolution of the Earth, McGraw-Hill, N.Y., 1981.

Dubos, René, The Unseen World, Oxford University Press, N.Y., N.Y., 1962.

Durant, W.J., Mansions of Philosophy, Garden City Publishing Co., N.Y., 1934.

Gosling, J.T., Waves in the Solar Wind, Scientific American, March, 1977.

Gould, Stephen J., The Panda's Thumb, Norton, N.Y., 1980.

Hoffding, H., The Problems of Philosophy, Macmillan Co., N.Y., 1906.

Horosz, William, Escape from Destiny, Chas. C. Thomas, Springfield, Ill., 1967.

James, William, Essays in Pragmatism, Hafner Publishing Co., N.Y., 1948.

Jaspers, Karl, The Great Philosophers, Harcourt, Brace and World, N.Y., 1962.

Laplan, A., The New World Philosophy, Random House, N.Y., 1961.

Lee, O., Existence and Inquiry, University of Chicago Press, Illinois, 1960.

Mc Laughlin, John C., The Tree of Animal Life, Dodd-Mead, N.Y., 1981.

Nietztche, F.W., Beyond Good and Evil, Boni and Liveright, N.Y., 1885.

Passmore, John, A Hundred Years of Philosophy, Basic Books, Inc., N.Y., 1966.

Persinger, M., ELF and VLF Electromagnetic Field Effects, Plenum Press, N.Y., 1974.

Pressman, A.S., Electromagnetic Fields and Life, Plenum Press, N.Y., 1970.

Russell, B.A.W., Problems of Philosophy, H. Holt Co., N.Y., 1912.

Vita

John Gibson Elliott was born in Chicago, Illinois. The son of Dr. John Adair Elliott and Mary Gibson Elliott. He entered the University of Oklahoma in 1963. Here he received his Bachelor's degree with a major in philosophy. In 1970 from the same University he received the Master of Liberal Studies degree. In 1974 he received the degree of Master of Science from Wright State University, Dayton, Ohio. In 1980 he received his Doctor of Arts degree.

He has been associated with the U.S. Government as a representative of NASA and the USAF. At Sinclair Community College, Dayton, Ohio, he taught social science. His hobbies are gardening and music.

NOTES

NOTES

NOTES